About the Marine Sanctuaries Conservation Series

The National Oceanic and Atmospheric Administration's Office of National Marine Sanctuaries (ONMS) administers the National Marine Sanctuary Program. Its mission is to identify, designate, protect and manage the ecological, recreational, research, educational, historical, and aesthetic resources and qualities of nationally significant coastal and marine areas. The existing marine sanctuaries differ widely in their natural and historical resources and include nearshore and open ocean areas ranging in size from less than one to over 5,000 square miles. Protected habitats include rocky coasts, kelp forests, coral reefs, sea grass beds, estuarine habitats, hard and soft bottom habitats, segments of whale migration routes, and shipwrecks.

Because of considerable differences in settings, resources, and threats, each marine sanctuary has a tailored management plan. Conservation, education, research, monitoring and enforcement programs vary accordingly. The integration of these programs is fundamental to marine protected area management. The Marine Sanctuaries Conservation Series reflects and supports this integration by providing a forum for publication and discussion of the complex issues currently facing the National Marine Sanctuary Program. Topics of published reports vary substantially and may include descriptions of educational programs, discussions on resource management issues, and results of scientific research and monitoring projects. The series facilitates integration of natural sciences, socioeconomic and cultural sciences, education, and policy development to accomplish the diverse needs of NOAA's resource protection mandate.

Normalization and Characterization of Multibeam Backscatter: Koitlah Point to Point of the Arches, Olympic Coast National Marine Sanctuary Survey - HMPR-115-2004-03

Steven S. Intelmann[1], Jonathan Beaudoin[2] and Guy R. Cochrane[3]

[1]Olympic Coast National Marine Sanctuary, NOAA
[2]Ocean Mapping Group, University of New Brunswick
[3]Coastal and Marine Geology Program, USGS

U.S. Department of Commerce
Carlos M. Gutierrez, Secretary

National Oceanic and Atmospheric Administration
VADM Conrad C. Lautenbacher, Jr. (USN-ret.)
Under Secretary of Commerce for Oceans and Atmosphere

National Ocean Service
John H. Dunnigan, Assistant Administrator

Office of National Marine Sanctuaries
Daniel J. Basta, Director

Silver Spring, Maryland
March 2006

DISCLAIMER

Report content does not necessarily reflect the views and policies of the Office of National Marine Sanctuaries or the National Oceanic and Atmospheric Administration, nor does the mention of trade names or commercial products constitute endorsement or recommendation for use.

REPORT AVAILABILITY

Electronic copies of this report may be downloaded from the National Marine Sanctuaries Program web site at www.sanctuaries.nos.noaa.gov. Hard copies may be available from the following address:

National Oceanic and Atmospheric Administration
Office of National Marine Sanctuaries
SSMC4, N/ORM62
1305 East-West Highway
Silver Spring, MD 20910

COVER

NOAA Ship *RAINIER* transiting between Sitka and Juneau, Alaska. The ship is designed and outfitted primarily for conducting hydrographic surveys in support of nautical charting. Photo credit Charles Doxley.

SUGGESTED CITATION

CONTACT

Steven S. Intelmann
Habitat Mapping Specialist
NOAA/National Marine Sanctuaries Program
N/ORM 6X26
115 E. Railroad Avenue, Suite 301
Port Angeles, WA 98362
(360) 457-6622 X22
steve.intelmann@noaa.gov

ABSTRACT

Through a partnership between the National Oceanic and Atmospheric Administration's (NOAA) Office of Coast Survey (OCS), NOAA's National Marine Sanctuary Program (NMSP), and NOAA's Office of Marine and Aviation Operations (OMAO), high resolution bathymetry (HRB) was collected on various opportunistic occasions during the months of October from 2001-2004 in the Olympic Coast National Marine Sanctuary (OCNMS). These particular survey operations were conducted aboard the NOAA ship *RAINIER* using a variety of multibeam echosounders suitable for the various regions of the sanctuary that were surveyed. Backscatter was derived from the Reson shallow water multibeam echosounders using custom software developed by researchers at the University of New Brunswick (Fredericton, Canada), for an area in the OCNMS, near Cape Flattery from Koitlah Point to Point of the Arches, and mosaiced at 1-meter pixel resolution. This process of normalizing the backscatter imagery significantly reduced the post-processing validation efforts that are required for the characterization effort. Textural classification of the sonar imagery suggests that nearly 58 percent of the seafloor in this area is covered by soft substrates such as mud or silt, 19 percent of the area is comprised of mixed sediment including cobbles, pebbles, gravel and boulders mixed with soft substrate, and over 23 percent of the total area is characterized by hard, complex rocky bottom. Video from a towed camera sled, bathymetry data, sediment samples, and the backscatter have been integrated to describe geological and biological aspects of habitat. Polygon features have also been created and attributed with a hierarchical deep-water marine benthic classification scheme (Greene et al. 1999). The data can be used with geographic information system (GIS) software for display, query, and analysis.

KEY WORDS

Benthic, habitat mapping, sediment classification, multibeam backscatter normalization, textural analysis, Olympic Coast National Marine Sanctuary, essential fish habitat, groundtruthing, accuracy assessment

TABLE OF CONTENTS

LIST OF FIGURES AND TABLES

INTRODUCTION

Having an increased understanding of the distribution and abundance of sea floor substrates can be useful for supporting management, research, monitoring, and education within the national marine sanctuaries (Barr 2003) and can assist with a myriad of concerns ranging from submarine cable engineering projects to cultural resource identification and protection. Seabed characterization was recently described as providing an important contribution to the sustainable management of groundfish fisheries (Kassakian and Ostdahl 2005) and has been integral to the Pacific Coast groundfish management (PFMC 2004) for the past several years.

Located off the Washington coast, the Olympic Coast National Marine Sanctuary (OCNMS) is a highly productive area of more than 2,400 nm^2 in size and supports an extensive fishery, yet information describing the distribution of benthic habitats for many of the groundfish species along the Washington coast is often sparse, especially at useful spatial scales (Wang 2005). In 2002, the OCNMS determined benthic habitat characterization as being a major program priority for the site, and has since devoted significant base resources toward gaining an increased understanding of the benthic environment. Although several sources of historic sediment grab sampling existed within the area (Venkatarathnam and McManus 1973; Nittrouer 1978; Sternberg 1986; Reid et al. 2001), it became evident that medium-scale acoustic geological sampling could better provide an effective means for describing the marine habitat (Greene et al. 1999; Valentine et al. 2003), and that technological innovations such as side scan sonar and multibeam sonar could better assist with delineating ocean floor substrates (Mitchell and Hughes Clark 1994; Auster et al. 1999; Cochrane and Lafferty 2002; Huvenne et al. 2002; Dartnell and Gardner 2004).

OCNMS has since undertaken various ship based acoustic (Intelmann and Cochrane 2006) and air-borne optical remote-sensing surveys (Intelmann 2005) to provide information for delineating benthic habitats within its jurisdictional boundaries. Of particular note, through a partnership with the National Oceanic and Atmospheric Administration's (NOAA) Office of Coast Survey (OCS), and NOAA's Office of Marine and Aviation Operations (OMAO), high resolution bathymetry (HRB) was collected on various opportunistic occasions during the months of October from 2001-2004. These particular survey operations were conducted aboard the NOAA ship *RAINIER* using a variety of multibeam echosounders suitable for the various regions of the sanctuary that were surveyed.

Backscatter was derived from the Reson shallow water multibeam echosounders using custom software developed by researchers at the University of New Brunswick (Fredericton, Canada), for an area in the Olympic Coast National Marine Sanctuary (OCNMS), near Cape Flattery from Koitlah Point to Point of the Arches (Hydrographic Survey Sheet A), and mosaiced at 1-meter pixel resolution. Textural classification (Cochrane and Lafferty 2002) was used to classify the mosaic into three distinct bottom types. Video from a towed camera sled, bathymetry data, sedimentary samples, and the backscatter have been integrated to describe geological and biological aspects of habitat.

Polygon features have also been created and attributed with a hierarchical marine benthic classification scheme (Greene et al. 1999), and an accuracy assessment was used to quantify classification performance. The data can be used with geographic information system (GIS) software for display, query, and analysis.

SURVEY AREA

Approximately 110 km^2 of sea floor were surveyed on the Hydrographic Survey Sheet A, located in the general vicinity of Cape Flattery and bounded by coordinates 48° 14'42'' N, 124° 51'06''W, and 48° 25'14'' N, 124°37'50''W (Figure 1). Survey work on Sheet A occurred from October 1-7, 2001; October 5-7, 2002; October 1-7, 2003; and October 11-12, 2004. Water depth throughout the project area ranged from 0.5 to 201 meters.

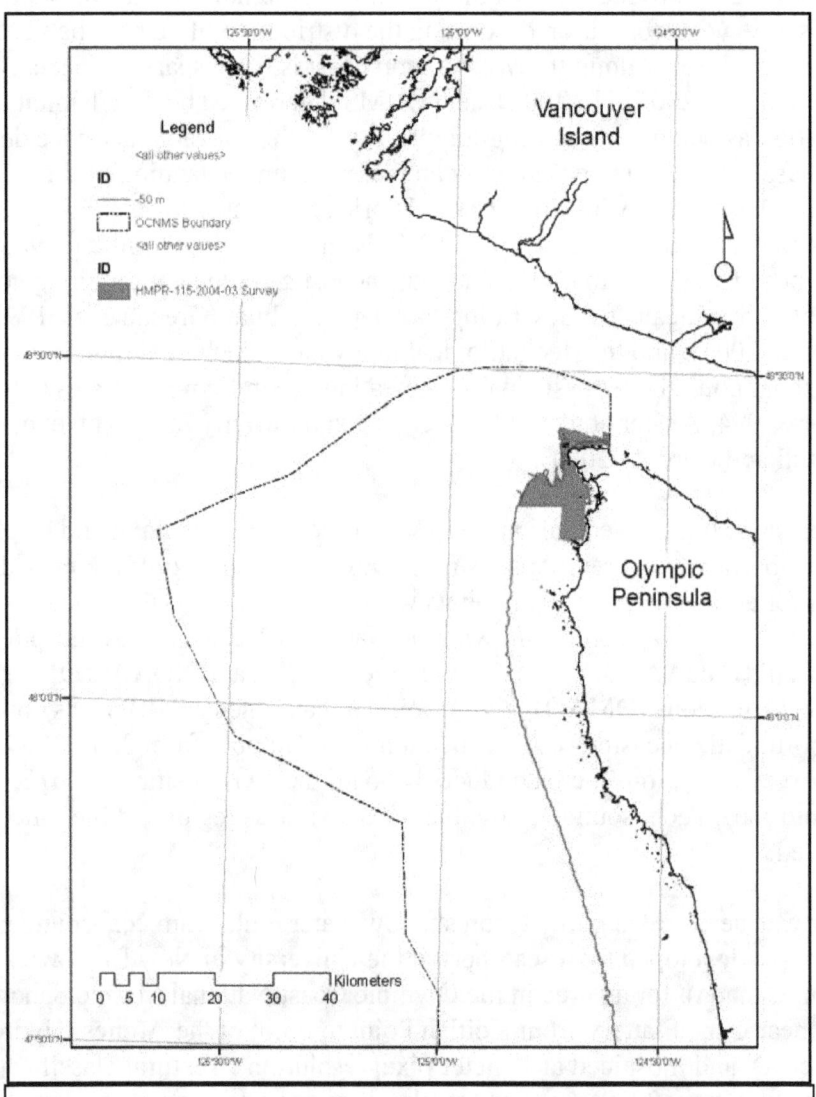

Figure 1. Koitlah Point to Point of the Arches (Hydrographic Survey Sheet A) survey footprint, shown with 50m bathymetry contour.

SONAR ACQUISITION

To maximize data acquisition effort, the NOAA ship *RAINIER*, measuring 70m in length, was used to simultaneously deploy multiple 9.8m survey launches (Figure 2). In general, survey launches were deployed at 0800 and retrieved at 1630 hours. Operations were frequently shortened or extended in response to changing weather and/or sea conditions, which varied greatly by day and year.

Each multibeam system was setup in a hull-mounted configuration as described in Table 1. After 2003, survey launch RA4 was outfitted with both an Reson 8125 (455 kHz) and an Elac 1180 (180 kHz) multibeam system. The 8125 is generally used to obtain full-bottom coverage in depths from 4-60 meters while the 1180 is used in depths from 50-300 meters, thus having the capability of both multibeam systems onboard increases the vessel's effective range of operation. Reson sonar data were logged in Extended Triton Format (XTF) using Isis Sonar (Triton Imaging International) with the "Full-New" side scan beam forming technique. This "new" process yields less noisy output by combining the bathymetry beams into two side scan beams where adjacent pairs of beams are then averaged and the brightest points of the averaged beams are ultimately selected (Reson 2003). Snippet packets were also logged from the 8101 systems during the 2003 and 2004 field operations but were not used in the mosaics since they were not captured in 2001 or 2002. Snippets are an intensity time series derived by composition from each narrow receive beam used in bottom detection. Snippet imagery is limited to only the bathymetric sector and is thus lost if bottom detection fails. Multiple samples per beam produces a better signal-noise ratio and additionally results in no multiple return artifacts. Elac sonar data were logged with HydroStar software.

Figure 2. Survey launch and retrieval operations aboard the NOAA ship *RAINIER*.

Vessel attitude and positioning for each of the launches was monitored with an TSS (Applanix) POS/MV 320 and logged in Isis Sonar. The POS/MV provides accurate attitude, heading, heave, position, and velocity data for each vessel. The ship's (RA) attitude was recorded using a TSS Meridian Attitude and Heading Reference System (MAHRS). The MAHRS is a heading reference instrument combined with a dynamic

3

motion sensor (DMS) to provide accurate heave, pitch and roll measurements. Aside from RA3, which was outfitted with an CSi MBX-2 differential GPS (DGPS) beacon, all other launches were equipped with Trimble Pro Beacon DGPS for survey line plan control. A Trimble DSM 212L DGPS was installed on the ship for line control and sonar positioning. Pro Beacon DGPS corrections were made using the Amphitrite Point base reference station, operating at 315 Khz in British Columbia, Canada.

To make necessary sound velocity corrections, and for Elac 1050D beam steering, either Seabird Seacat SBE 19 or SBE 19plus CTD-profilers were used to acquire information about physical properties of the water column that dictate ray tracing, or essentially how sound will travel through the water. Water level observations were acquired from the Neah Bay tidal station 9443090 and applied with zoned corrections. Line planning and project control were accomplished through Hypack marine positioning and surveying software. Vessel speed was targeted at 8 knots during sonar acquisition.

SONAR DATA PROCESSING AND IMAGE CLASSIFICATION

Production of acoustic backscatter imagery was done with software tools developed by the Ocean Mapping Group (OMG), University of New Brunswick (Beaudoin et al. 2002). The software corrects for variations in acoustic source level, receiver gain, and pulse length, in addition to modeling/removing the effects of transmitter and receiver beam patterns. The work described in Beaudoin (2002) only addresses variations in source level and receiver gain; the software has been upgraded to address the remaining aforementioned corrections since then. Additionally, the software performs an across-track signal normalization to minimize the variations due to the angular response of the seafloor. Beam pattern models were used for beam pattern normalization since it was not logistically possible to measure the beam pattern of each launch directly. Further, the acoustic backscatter derived from the OMG software is an approximation because a calibration of the receiver was not performed and an estimate had to be used to convert receiver digital data to physical units of pressure. This approximation would not be an issue with classification performance since it would only introduce a bulk shift in the acoustic backscatter values across the entire survey area, leaving relative differences between sediment regimes unaffected.

Commercially available software packages available for processing multibeam backscatter imagery perform only a rudimentary geo-registration through the use of a flat seafloor assumption. Such assumption introduces positioning errors that grow with the seafloor's deviation from an ideal flat surface. Thus geometric corrections were performed using the OMG software that allowed for the more accurate geo-registration of the acoustic backscatter; this is performed using the corresponding bathymetric profile derived from the multibeam system. Because no bathymetric data is collected beyond the angular sector of the multibeam system, OMG software does not geo-register the acoustic backscatter data beyond this point (red lines in Figure 3).

4

Figure 3. Comparison of image mosaicing produced by commercially available software (left) with the same survey line processed using custom software developed by the UNB OMG (right). Red line depicts coverage of bathymetric data, which is used as a stencil by OMG software.

The process and need for normalizing multibeam backscatter data, as is possible with the OMG Reson processing routine, becomes paramount when a goal of data acquisition is to create classified habitat maps. The signal return strength from any sonar system is modulated by its propagation through the oceanic environment, its interaction with the seafloor, and by variable sonar system parameters. In order to be used for seafloor characterization, sonar echo strength data must be corrected for all variations, leaving the seafloor's backscattering strength as the sole source of signal strength variation. Oceanographic variations are readily modeled and removed through time-varying gain functions that are routinely implemented in the sonar receiver hardware (as is the case with the Reson systems discussed herein). Signal strength variations due to sonar system parameters such as source level, receiver gain, pulse length, and transmitter/receiver beam patterns must then be addressed. Provided that sufficient documentation of the sonar system parameters is available, these variations can be removed from the return signal. The final hurdle involves normalizing the return signal to compensate for signal strength variation due to imaging geometry. Because multibeam sonars ensonify the seafloor over a large swath, signals returning from below the vessel are typically much stronger than signals returning from the outer portions of the swath, as evident by the white strip in the left image of Figure 3. This large variation must be normalized; otherwise seabed characterization algorithms will characterize the seabed below the vessel as being different from the seabed away from the vessel. In comparing OMG processed results to those derived from commercially available software (as in Figure 3), it is obvious that seafloor angular response and beam pattern normalization is required for seabed characterization algorithms to produce desirable results. For example, whilst examining the data below the vessel track in Figure 3, characterization algorithms would classify the white strip as being of a different sediment type than the remainder of the data. Clearly, this would be incorrect and the analyst would spend increased time correcting the algorithm's output. After processing and normalization using OMG software, the white strip no longer has the detrimental effect on the characterization algorithms, effectively leaving the analyst with much less post-processing validation work.

Backscatter was processed only from the Reson multibeam systems because the Elac data format does not preserve the parameters necessary for signal normalization (e.g. source level, gain, pulse length, etc). Backscatter data were individually normalized through separate process routines for each survey launch by year. An image stretch was applied to individual mosaics to maximize use of grey scale bandwidth. Processed mosaics from each survey launch were then imported to Imagine image processing software (Leica Geosystems) and merged into a single image file (Figure 4). The merged mosaic was then split into four geographic regions for more manageable image computation during the classification process. The four "regional" mosaics were converted to a raw binary format comprised of only grey-scale pixel intensities ranging from 0 to 255.

Several studies (Skohr 1991; Blondel 1996) have found the use of grey-level alone for assigning classification codes to side scan sonar imagery as being inadequate, and other studies (Blondel 1996; Cochrane and Lafferty 2002; Huvenne et al. 2002; Intelmann and Cochrane 2006) have successfully used various textural indices to more effectively classify side scan sonar data, thus we used a second-order textural analysis (Cochrane and Lafferty 2002) on the raw binary images to differentiate bottom types from the backscatter imagery. Using a co-occurrence matrix approach provides an alternative for classifying acoustic imagery, and has been found to more effectively assess the spatial relationship of pixel intensities from remote sensing data (Haralick 1973; Blondel 1996). Indices for homogeneity and entropy were calculated for each regional mosaic, thereby producing textural measures for roughness and organization.

For each of the four regionally split images, the original backscatter data (or DN image), entropy index, and the homogeneity index images were all geo-referenced and stacked into a single Imagine file, creating a "pseudo" multi-spectral image file. Using knowledge gained from video data, training signatures were manually digitized in the Imagine classification module for areas representing soft, mixed, and hard bottom types (see Appendix). Soft bottom included mud, sand or silt. Mixed bottom included a combination of soft bottom and/or sand, shell hash, gravel, pebble, or cobble. Areas of boulders or rock outcrops were classified as hard bottom. Using a maximum likelihood decision rule, a supervised classification procedure was independently run on each of the four "pseudo" multi-spectral images. The output resulted in four separate classified thematic raster images that were converted back to grey scale and again exported as raw binary files. Unclassified, misclassified (such as at nadir) or poorly classified data were manually reclassified in Adobe Photoshop by digital overlay and cross-referencing the thematic image with the original backscatter (DN) images. Each of the four cleaned grey

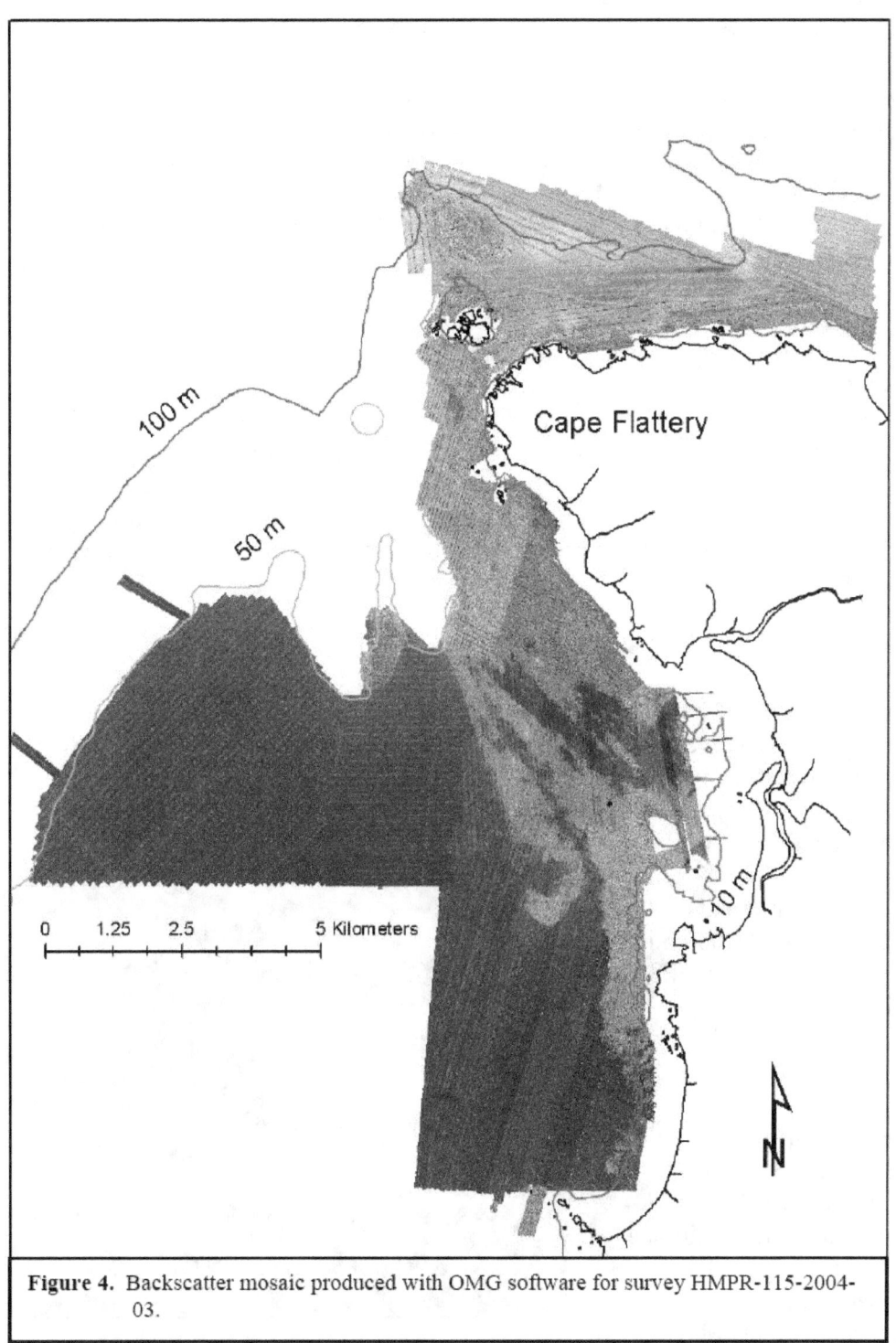

Figure 4. Backscatter mosaic produced with OMG software for survey HMPR-115-2004-03.

scale images were then imported back into Imagine and mosaiced into one classified image for the entire survey area.

The MajorityFilter command in ArcInfo (ESRI) was then used to filter the image to reduce the number of polygon features and essentially smooth the data through a 3x3 neighborhood analysis. The raster image was then converted to a polygon feature using ArcInfo. Video and sediment grab data were used as a validation tool to manually split polygons, where needed, and attach attributes describing macrohabitat according to Greene et al. (1999). Polygons features were manually added or removed where backscatter data quality was too poor to allow for reliable textural classification, such as acoustic shadows, or where features could be better discerned from overlay with a high-resolution shaded digital terrain model created from the multibeam sounding data.

Individual line cleaning and subset editing of the bathymetry data was accomplished using Hips software (Caris). Additional area-based edits were made with Fledermaus (Interactive Visualization Systems) to insure more thorough data cleaning. The bathymetry data was used to create two separate grids representing seafloor slope and complexity, the latter through a standard deviation neighborhood analysis (3x3). Both the slope and complexity grids were converted to polygon features and reclassified according to Greene et al. (1999), although classes were modified to better represent the physical characteristics of this particular survey area. Slope and complexity values were then filtered and joined with the bottom polygon feature through a UNION process.

GROUNDTRUTHING

In August 2005, the NOAA vessel *Tatoosh* was used to tow a Taras camera sled (Figure 5) to acquire underwater videography for validating the sonar imagery. The tow sled was configured with a Deep Sea Power & Light SeaCam, SeaLite, and dual SeaLasers (mounted with 10cm spacing), TriTech 200 kHz altimeter, and an Applied Acoustic micro beacon. The video was sent through Falmat Xtreme Green video cable and captured with a Sony GV-D1000 mini-DV recorder. A Sea-Trak GPS overlay was used to dub positioning information onto the video.

Since no acoustic positioning was available, the tow sled was drifted directly below the A-frame to minimize positional offset from the vessel's DGPS antennae. In instances where significant currents were

Figure 5. Video sled deployment for groundtruthing.

encountered, effort was made to actively tow the sled with the current to reduce layback. The USSEABED project, a database providing information on sediment and rock distributions in the waters off the United States (Reid et al. 2001), was also queried and provided 94 samples to describe sedimentology within the survey area. Video transects and USSEABED sample locations are shown in Figure 6.

An accuracy assessment of the texture classification was accomplished by converting the navigation data from the video sled into an xyz text file, where z represents bottom hardness (hard, mixed, soft). The texture classification image was then compared to the video observations to calculate overall accuracy and Kappa statistics (Table 4).

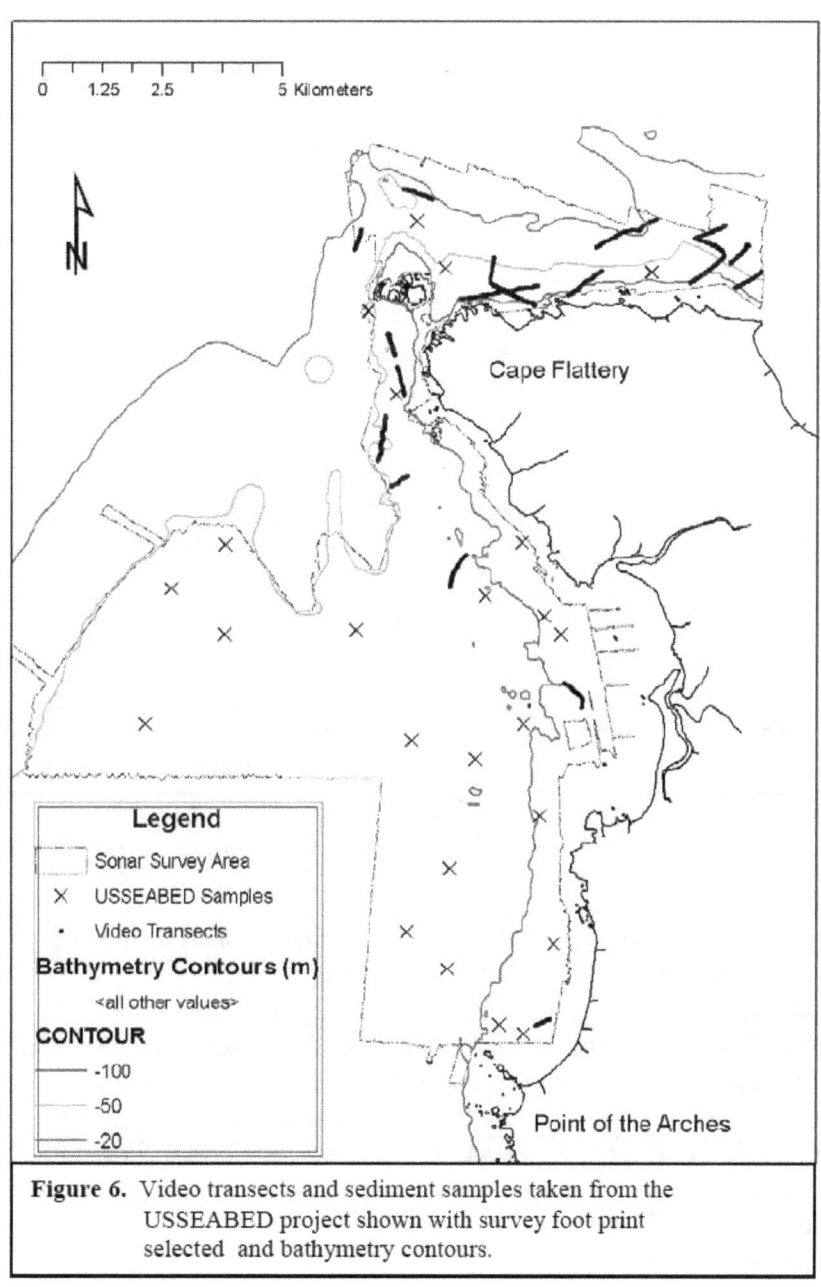

Figure 6. Video transects and sediment samples taken from the USSEABED project shown with survey foot print selected and bathymetry contours.

SURVEY RESULTS AND INTERPRETATION

Over 98 km^2 near Cape Flattery, from Koitlah Point to Point of the Arches, were surveyed with Reson SeaBat multibeam echosounders, the remainder using L-3 Seabeam Elac multibeam systems (Table 1). Approximately 945 linear km of track lines were collected throughout the effort, yielding more than 119 hours of logged sonar data. An additional twenty hours of video were captured as groundtruthing validation.

Table 1. Data acquisition effort by year and survey launch. Acquisition time refers to actual logged data, derived from start and stop times in XTF headers. Linear kilometers of track lines was calculated with NOAA HSTP's Pydro software.

Year	Survey Launch	MB Sounder[1]	Acq. Time	Linear Km	Area[2]
2001	RA1	8101	21:59:08	151.8	27.1
	RA6	8101	18:25:19	203.0	
2002	RA4	8125	11:16:08	83.3	19.4
	RA5	8101	10:36:41	86.8	
2003	RA3	8101	15:11:42	123.2	42.6
	RA4	8125	14:56:16	97.8	
	RA5	8101	17:52:13	130.6	
2004	RA4	1180	2:28:59	18.7	20.4
	RA5	8101	2:45:05	18.6	
	RA	1050D	3:46:29	30.8	
Total			**119:18:00**	**944.6**	**109.5**

[1]Multibeam systems used for the survey were either a Reson SeaBat 8101, Reson SeaBat 8125, L-3 Seabeam Elac 1050D, or L-3 Seabeam Elac 1180. [2]Area calculations are presented as a combination of survey platforms for each year.

Megahabitat (Greene et al. 1999) was defined as continental shelf throughout the entire survey area. Textural classification of the sonar imagery suggests that nearly 58 percent of the seafloor in the survey area near Cape Flattery, from Koitlah Point to Point of the Arches, is covered by soft substrates such as mud or silt, 19 percent of the area is comprised of mixed sediment including cobbles, pebbles, gravel and boulders mixed with soft substrate, and over 23 percent of the total area is characterized by hard complex rocky bottom (Table 2). Based on an accuracy assessment report of bottom hardness, overall classification accuracy was 86.11 percent with an overall Kappa statistic of 0.7596 (Table 4).

Towed camera and visual observations at low tide indicate the majority of the hard rocky outcrops found in the nearshore resemble those same middle-upper Eocene marine sedimentary rocks identified directly onshore (Dragovich et al. 2002). Several geologic forces were evident in the sonar imagery including areas of tilted, differentially eroded bedrock. The two parallel, northwest-trending fault strands located offshore of Makah Bay are further evidence of the anticlinal folding and thrust faulting that occur in this geologic collision zone (McCrory et al. 2004). Multiple areas were characterized as having large distinct sediment waves (Table 3). One particularly notable area of sediment waves identified in the video imagery, located approximately 1 km offshore of Chibahdehl Rocks, consisted of large mega-ripples likely attributed to the high tidal currents that occur in that particular region of the Strait of Juan de Fuca.

Table 2. Bottom hardness classified from HMPR-115-2004-03 survey data. Bottom induration codes (Greene et al. 1999) are provided by area in square meters, and area as percentage of total mapped area.

Bottom_ID	Descriptor	Square m	Percentage
h	Hard	22,900,464	23.2
m	mixed	18,670,687	18.9
s	Soft	57,143,375	57.9

Table 3. Habitat types classified from HMPR-115-2004-03 survey data. Habitat codes are provided per Greene et al. (1999) and are presented by area in square meters, and area as percentage of total mapped area. Code concatenation for slope, complexity, and microhabitat have been omitted from table for brevity.

Habitat Code	Descriptor	Square m	Percentage
Ss_u	Shelf soft unconsolidated	52,287,128.6	53.97
She	Shelf hard exposure (bedrock)	19,642,359.7	19.90
Sm	Shelf mixed	18,360,151.9	18.60
Ss	Shelf soft	4,853,227.9	4.92
Shd_d	Shelf hard deformed differential erosion	3,225,211.0	3.27
Smw_r	Shelf mixed waves ripples	283,816.8	0.29
Sh	Shelf hard	32,893.2	0.03
Smw	Shelf mixed sediment waves	26,718.5	0.03
Ssw	Shelf soft waves	3018.2	<0.01

Table 4. Contingency matrix, accuracy totals, and Kappa Statistics for accuracy assessment analysis.

Error Matrix

Classified Data	no_data	Reference Data M	h	s
no_data	0	1	0	0
m	0	**15658**	802	2094
h	0	427	**7035**	584
s	0	16	274	**3333**
Column Total	0	16102	8111	6011

Accuracy Totals

Class Name	Reference Totals	Classified Totals	Number Correct	Producers Accuracy	Users Accuracy
No_Data	0	1	0	--------	--------
m	16102	18554	15658	97.24%	84.39%
h	8111	8046	7035	86.73%	87.43%
s	6011	3623	3333	55.45%	92.00%
Totals	30224	30224	26026		
Overall Accuracy	**86.11%**				

Kappa Statistics

Overall Kappa	0.7596	Class Name	Kappa
		1	0.6659
		2	0.8203
		3	0.9001

Microhabitat and presence of certain biologic attributes were also populated into the polygon features, but were strictly limited to areas where video groundtruthing occurred and where the sea floor was clearly visible in the footage. Figure 7 provides a graphical representation of the habitat characterization including the codes for microhabitat and biologic attributes.

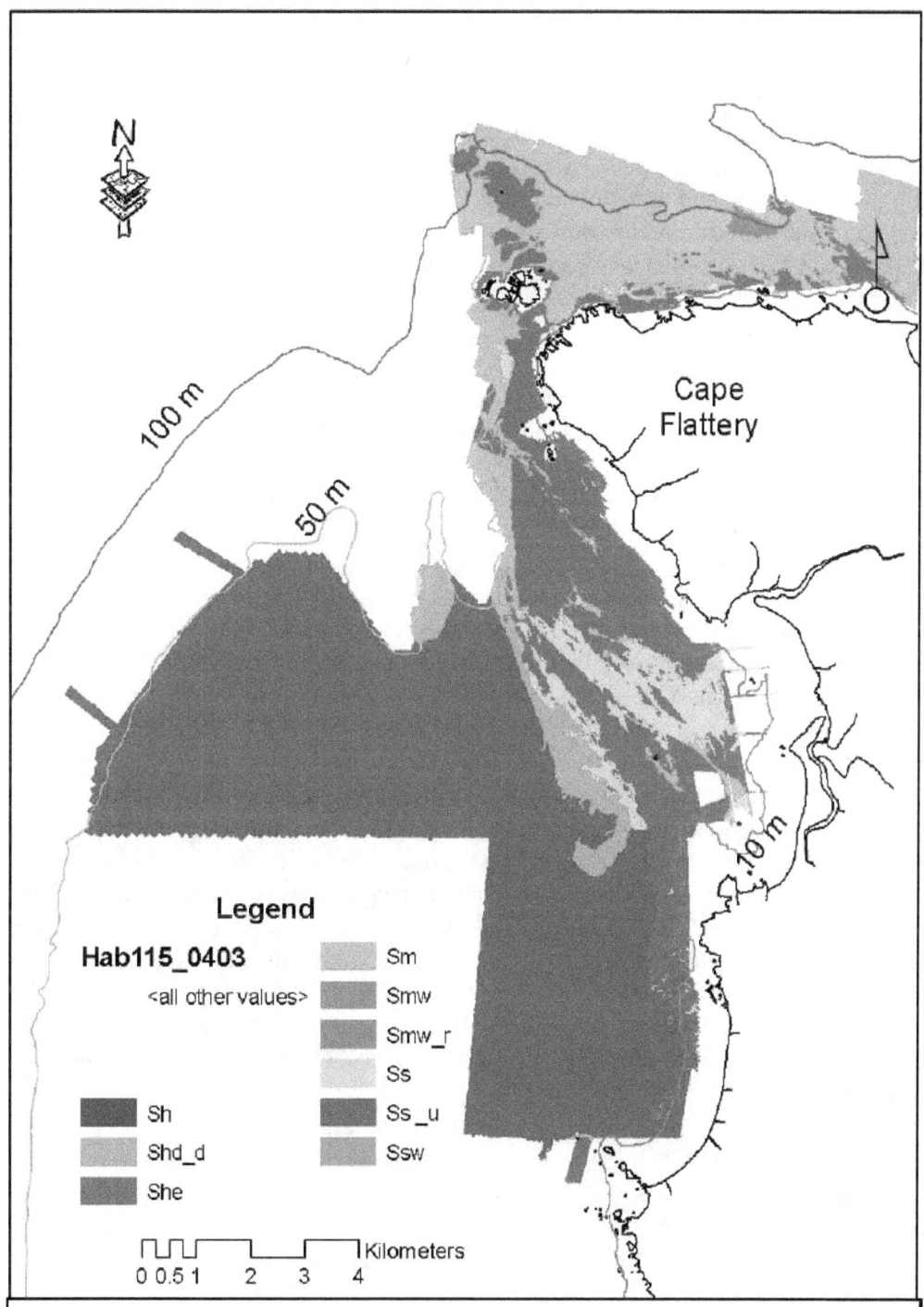

Figure 7. Habitat classification for survey HMPR-115-2004-03. As listed in Table 3, codes in the image only indicate a habitat class concatenation of megahabitat, bottom induration, meso-scale identifier, and modifier id. Other fields have been omitted for brevity.

DISCUSSION AND CONCLUSIONS

The assistance provided by NOAA's OCS, Hydrographic Survey Division, permitted OCNMS to sporadically acquire a significant amount of hydrographic survey data over a period of four years, at minimal cost to the program. The OCS is responsible for creating, maintaining, and updating nautical charts of U.S. waters, with critical survey areas being determined from a variety of factors including marine traffic patterns and inadequacy of prior surveys. The combined priority for creating contemporary high-resolution sea floor maps of the nation's marine sanctuaries, and the need for updating the nautical charts along the western coast of Washington, created a unique opportunity for an intra-agency cooperation within NOAA between the NMSP, OCS, and OMAO. The expertise and capabilities of the NOAA ship *RAINIER* proved to be an extremely valuable asset for advancing the sanctuary program's characterization efforts.

Additional sea days by the *RAINIER,* her sister ship *FAIRWEATHER*, or the *HI'IALAKAI* would be of great value to the OCNMS' continued high priority habitat mapping program. Furthermore, a regular commitment by OMAO to support the entire NMSP's mapping efforts through the fleet allocation process would help to accomplish NOAA's ecosystem mission goal, which specifically states mapping and characterizing previously uncharted habitats as being a key foci in the 5-year research plan (NOAA 2005). Continued OMAO support would also provide critical support for NOAA's overall mission (NOAA 2005). Even though the overall assets of the *RAINIER* are not entirely compatible with the mapping goals of the OCNMS, as the multibeam system installed on the ship itself is not capable of preserving the parameters necessary for signal normalization (e.g. source level, gain, pulse length, etc.), the systems available on the various launches were capable of providing useful data for creating quality benthic habitat maps. However, even the multibeam data acquired by the ship's lower frequency Elac multibeam systems provided useful data for creating digital terrain models. Such data, for example, can be extremely valuable as a reconnaissance tool for deep-water side scan surveys, especially in the canyon areas (Intelmann and Cochrane 2006). The deep-water multibeam systems installed on the NOAA Ships *FAIRWEATHER* and *HI'IALAKAI,* in addition to those installed on their survey launches, are all capable of capturing the necessary parameters for normalizing backscatter data, and as such, those platforms could potentially provide even more useful data for characterization efforts than the *RAINIER*. The sanctuary's continued high priority for describing benthic habitats and the close physical home port locality of the NOAA hydrographic survey vessels, especially the *RAINIER*, provide a prime opportunity for cross program collaboration that should be utilized to an even greater degree.

As previously stated, the process and need for normalizing multibeam backscatter data, as is possible with the OMG Reson processing routine, becomes paramount when a goal of data acquisition is to create classified habitat maps. A benefit of using backscatter, as opposed to traditional side scan imagery, for seafloor characterization is the generally more precise positioning achievable through a hull-mounted configuration. The data are also precisely geo-coordinated with the multibeam sounding data and thus avoid the

14

various challenges associated with the geo-positioning of devices such as towed side scan sonar. However, a drawback of multibeam backscatter is that because the instrument is hull-mounted, image resolution will typically decline with increasing water depths. Inherent transducer design and the ability to tow a side scan sonar closer to the seafloor usually results in more pronounced image textural properties, which are important for calculating various textural indices, such as homogeneity and entropy as used in our characterization process. Also, since multibeam swath coverage depends directly on water depth, data acquisition productivity will decrease in shallower water (Kamoshita et al. 2005). Survey productivity in shallow water can be increased by removing the angular dependence of multibeam, as the case with side scan sonar or interferometric bathymetric methods.

The accuracy assessment matrix provides useful information pertaining to the textural classification performance; however, in this survey the calculated statistics are probably lower than reality due to the lack of precise acoustic positioning on the towed camera sled and possible changes in the seafloor sediment regime between the time of acoustic imaging and the time of video groundtruthing. In some areas nearly four years separated the two independent survey efforts. With the highly variable tidal currents existing across some of the survey area it is highly probable that sediments shifted over time, thereby introducing error not necessarily attributable to the classification performance. Moreover, a reliable accuracy assessment should be based on the best possible reference data. In our groundtruthing survey, the reference data were derived from a video sled using positioning based on the vessel rather than the actual video sled itself as no acoustic tracking system was available. In this survey, reference data will have a minimum 6 meter error because that was the DGPS antennae offset from the A-frame that served as the camera sled tow point. Further, the assumed position of the reference data would be more accurate in areas that experienced less influence from the current and in the areas of shallower water throughout the survey area, where less cable was deployed, ultimately translating to less potential layback error.

In examining the error matrix (Table 4), it is important to assess the diagonal cells, as they indicate the observations classified correctly according to the reference data (i.e., video groundtruthing). Observations deviating from the diagonal indicate misclassified data for each category. The overall classification accuracy statistic is useful information but should be interpreted with caution because it simply describes the number of incorrect observations divided by the number correct. The user's accuracy essentially provides a measure for classification performance in the field by class, and the producer's accuracy depicts how accurately the analyst classified the image by class. The Kappa statistic provides a more powerful multivariate descriptor, incorporating both the diagonal observations (as used in the overall accuracy) and the off diagonal observations. The Kappa statistic essentially indicates the proportionate reduction in error generated by the classification process as opposed to a completely random generated classification (Leica Geosystems 2003). For this work, the overall Kappa statistic of 0.7596 implies that the classification process is avoiding nearly 76 percent of the errors that would be associated with a completely random classification process (Congalton 1991).

ACKNOWLEDGEMENTS

The authors would like to thank the crew of the NOAA ship *RAINIER* for their excellent performance with data acquisition and cursory line cleaning of the multibeam sounding data; Andy Palmer and Mike Levine for skippering the R/V *Tatoosh* and assistance with the groundtruthing survey; and Rich Littleton for assisting with the video sled deployment. The manuscript benefited from additional review by CDR Guy Noll, NOAA Corps, Commanding Officer, NOAA ship *RAINIER*, and edits by Kathy Dalton.

REFERENCES

Auster, P.J. Michalapoulos, R. Robertson, P.C. Valentine, K. Joy, and V. Cross. 1999. Use of acoustic methods for classification and monitoring of seafloor habitat Complexity: description of approaches. Pages 186-197 in: N.W. Munro and J.H.M. Wilison (eds.) Linking Protected Areas With Working Landscapes. Sciense and Management of Protected Areas Association, Wolfsville, Nova Scotia.

Barr, B. 2003. US Geological Survey NOAA/National Marine Sanctuary Program's seabed mapping initiative. 2002-2003 National Marine Sanctuary Annual Report.

Beaudoin, J., Hughes Clarke, J.E., Van den Ameele, E. and Gardner, J., 2002, Geometric and radiometric correction of multibeam backscatter derived from Reson 8101 systems: Canadian Hydrographic Conference 2002 Proceedings (CDROM), Toronto, Canada.

Blondel. P. 1996. Segmentation of the Mid-Atlantic Ridge south of the Azores, based on acoustic classification of TOBI data. In: MacLeod, C.J. Tyler, P.A. Walker, C.L. (Eds.), Tectonic, Magmatic, Hydrothermal and Biological Segmentation of Mid-Ocean Ridges. Geological Society Special Publication No. 118. Boulder, CO. pp 17-28.

Cochrane, G.R., and K.D. Lafferty. 2002. Use of acoustic classification of sidescan sonar data for mapping benthic habitat in the Northern Channel Islands, California. Continental Shelf Research 22: 683-690.

Congalton, R. 1991. A review of assessing the accuracy of classifications of remotely sensed data. Remote Sensing of Environment 37: 35-46.

Dartnell, P., and J.V. Gardner. 2004. Predicting seafloor facies from multibeam bathymetry and backscatter data. Photogrammetric Engineering & Remote Sensing. 70(9): 1081-1091.

Dragovich J.D., R.L. Logan, H.W Schasse, T.J. Walsh, W.S. Lingley Jr., D.K. Norman, W.J. Gerstel, T.J. Lapen, J.E. Schuster, and K.D. Meyers. 2002. Geologic map of Washington-Northwest Quadrant. Washington Division of Geology and Earth Resources. Geologic Map GM-50.

Greene, H.G., M.M. Yoklavich, R.M. Starr, V.M. O'Connell, W.W. Wakefield, D.E. Sullivan,J.E. McRea, Jr., G.M. Cailliet. 1999. A classification scheme for deep seafloor habitats. Oceanologica Acta. 22(6):663.

Haralick, R.M., K. Shanmugan, and R. Dinstein. 1973. Textural features for image classification. IEEE Transactions Systems, Man, and Cybernetics SMC3, 610-621.

Huvenne, V.A.I., Ph. Blondel, and J.P. Henriet. 2002. Textural analyses of sidescan sonar imagery from two mound provinces in the Porcupine Seabight. Marine Geology (189):323-341.

Intelmann, S.S. 2005. Hydrographic and topographic lidar acquisition. Northwest Coast, Washington. Neah Bay to Cape Alava. Unpublished Data.

Intelmann, S.S. and G.R. Cochrane. 2006. Benthic Habitat Mapping in the Olympic Coast National Marine Sanctuary: Classification of side scan sonar data from survey HMPR-108-2002-01: Version I. Marine Sanctuaries Conservation Series MSD-06-01. U.S. Department of Commerce, National Oceanic and Atmospheric Administration, Marine Sanctuaries Division, Silver Spring, MD. 13pp.

Kamoshita, T., Y. Sato, and T. Komatsu. 2005. Hydro-acoustic survey scheme for sea-bottom ecology mapping. Seatechnology 46(6): 39-43.

Kassakian J. and M. Ostdahl. 2005. Memorandum to Washington State Policy Working Group. Improving the sustainability of fisheries off Washington's outer coast through benthic habitat mapping and characterization. November 3, 2005.

Leica Geosystems GIS & Mapping, LLC. 2003. Erdas Field Guide. 7th Edition. Atlanta, Georgia. 261-262.

McCrory, P.A., S.C. Wolf, S.S. Intelmann, W.W. Danforth, R.J. Weldon, J.L. Blair. 2004. Quaternary tectonism in a collision zone, Northwest Washington. Eos Trans. AGU, 85(47), Fall Meet. Suppl., Abstract T33C-1391.

Mitchell, N.C., and J.E. Hughes Clark. 1994. Classification of seafloor geology using multibeam sonar data from the Scotian shelf. Marine Geology 121: 143-160.

Nittrouer, C.A., 1978. The process of detrital sediment accumulation in a continental shelf environment: an examination of the Washington shelf: PhD Thesis, University of Washington, Seattle, WA. 243p.

NOAA 2005. Research in NOAA: Toward understanding and predicting earth's environment. A five-year plan: fiscal years 2005-2009. US Department of Commerce NOAA. January 2005. 60pp.

NOAA 2005. New priorities for the 21st century – NOAA's strategic plan. Updated for FY 2006-FY2011. US Department of Commerce NOAA. April 2005. 24 pp.

Pacific Fishery Management Council, 2004. Pacific coast groundfish fishery management plan for the California, Oregon, and Washington groundfish fishery as amended through Amendment 17. Pacific Fishery Management Council. Portland, OR. 145 pp.

Reid, J. A., Jenkins, C., Field, M. E., Gardner, J. V. and Box, C. E. 2001. USSEABED: defining the surficial geology of the continental shelf through data integration and fuzzy set theory. Geological Society of America Annual Meeting, Boston, MA. Abstracts with Programs 33:A106.

Reson 2003. SeaBat 8101 Multibeam echosounder system operator's manual. Version 3.01.

Skohr, M.E. 1991. Evaluation of second-order texture parameters for sea ice classification from radar images. Journal of Geophysical Research. 96: 10625-10640.

Sternberg, R.W. 1986. Transport and accumulation of river-derived sediment on the Washington continental shelf, USA. Journal of the Geological Society of London, v. 143, p. 945-956.

Valentine, P.C., G.R. Cochrane, and K.M. Scanlon. 2003. Mapping the seabed and habitats in National Marine Sanctuaries. Marine Technology Society. 37(1): 10-17.

Venkatarathnam, K., and McManus, D.A. 1973. Origin and distribution of sands and gravels on the northern continental shelf off Washington. Journal of Sedimentary Petrology, v. 43, p. 799-811.

Wang, S.S-E. 2005. Groundfish habitat associations from video survey with a submersible off the Washington Coast. Master's Thesis. University of Washington. 110pp.

APPENDIX

Still frames extracted from video for representative areas classified as **soft** bottom

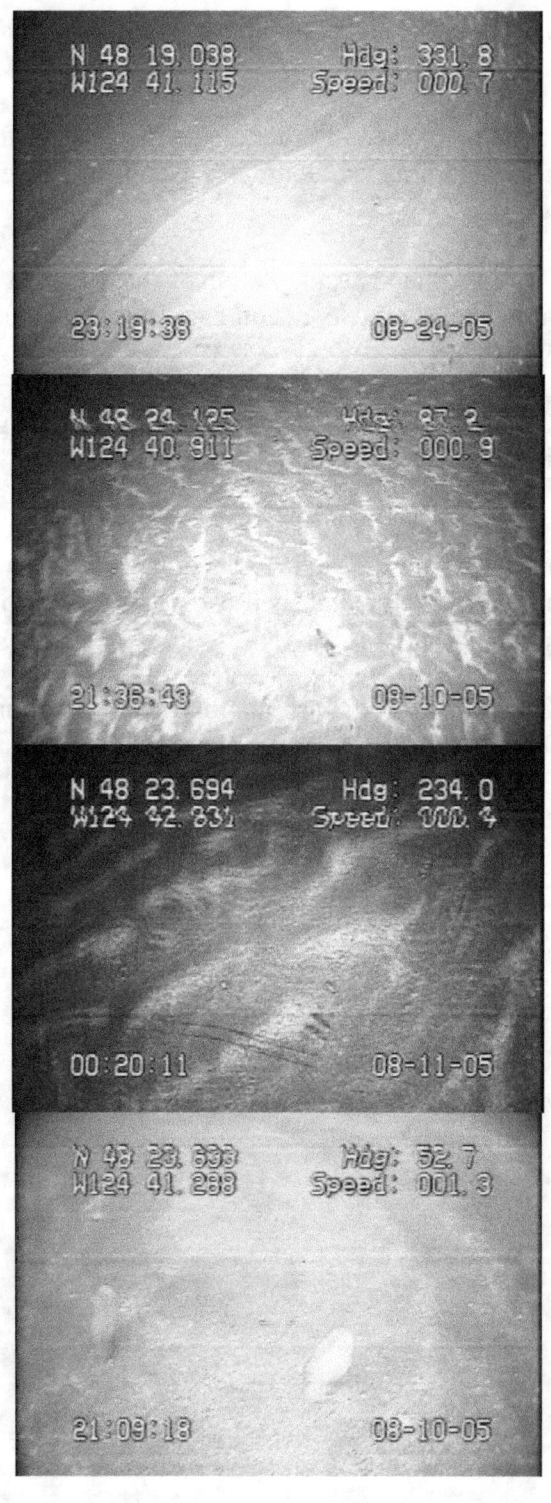

Still frames extracted from video for representative areas classified as **mixed** bottom.

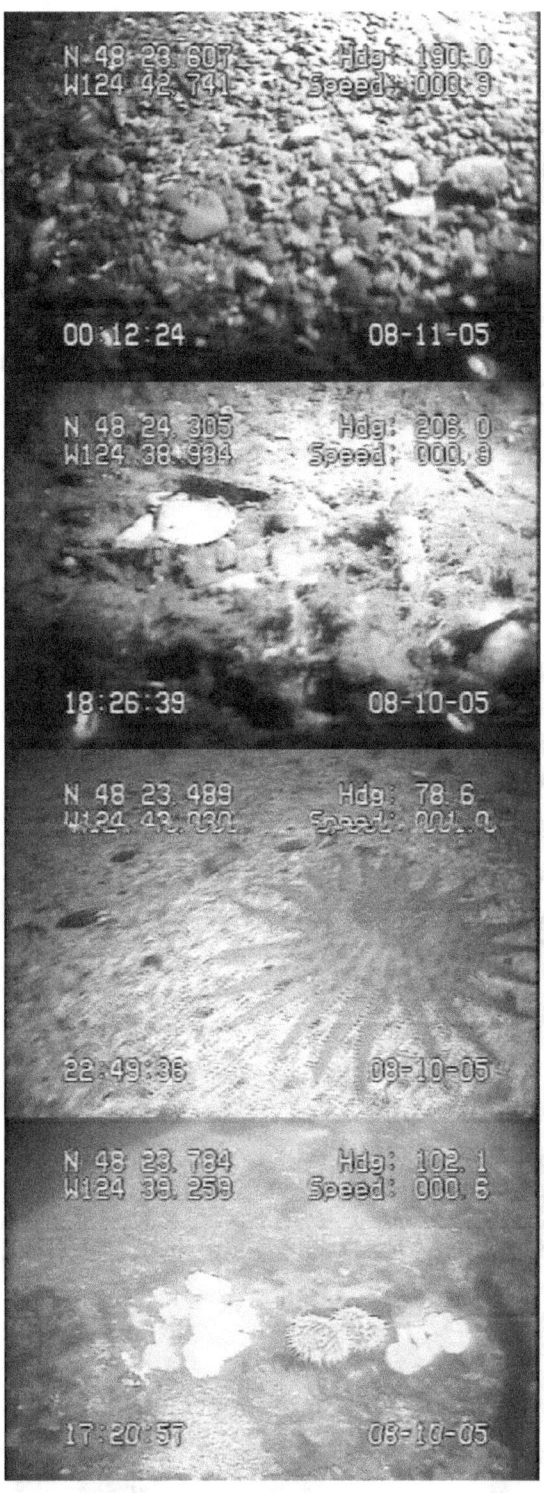

Still frames extracted from video for representative areas classified as **hard** bottom.

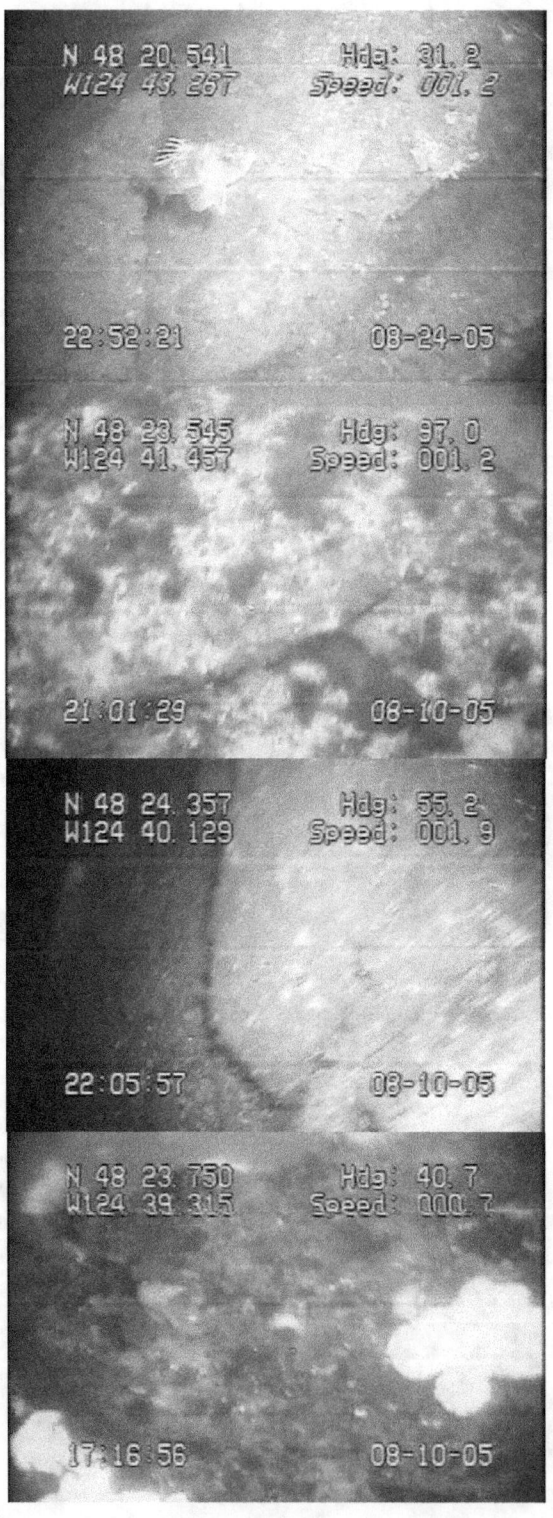

ONMS CONSERVATION SERIES PUBLICATIONS

To date, the following reports have been published in the Marine Sanctuaries Conservation Series. All publications are available on the Office of National Marine Sanctuaries website (http://www.sanctuaries.noaa.gov/).

Developing Alternatives for Optimal Representation of Seafloor Habitats and Associated Communities in Stellwagen Bank National Marine Sanctuary (ONMS-06-02)

Benthic Habitat Mapping in the Olympic Coast National Marine Sanctuary (ONMS-06-01)

Channel Islands Deep Water Monitoring Plan Development Workshop Report (ONMS-05-05)

Movement of yellowtail snapper (*Ocyurus chrysurus* Block 1790) and black grouper (*Mycteroperca bonaci* Poey 1860) in the northern Florida Keys National Marine Sanctuary as determined by acoustic telemetry (MSD-05-4)

The Impacts of Coastal Protection Structures in California's Monterey Bay National Marine Sanctuary (MSD-05-3)

An annotated bibliography of diet studies of fish of the southeast United States and Gray's Reef National Marine Sanctuary (MSD-05-2)

Noise Levels and Sources in the Stellwagen Bank National Marine Sanctuary and the St. Lawrence River Estuary (MSD-05-1)

Biogeographic Analysis of the Tortugas Ecological Reserve (MSD-04-1)

A Review of the Ecological Effectiveness of Subtidal Marine Reserves in Central California (MSD-04-2, MSD-04-3)

Pre-Construction Coral Survey of the M/V Wellwood Grounding Site (MSD-03-1)

Olympic Coast National Marine Sanctuary: Proceedings of the 1998 Research Workshop, Seattle, Washington (MSD-01-04)

Workshop on Marine Mammal Research & Monitoring in the National Marine Sanctuaries (MSD-01-03)

A Review of Marine Zones in the Monterey Bay National Marine Sanctuary (MSD-01-2)

Distribution and Sighting Frequency of Reef Fishes in the Florida Keys National Marine Sanctuary (MSD-01-1)

Flower Garden Banks National Marine Sanctuary: A Rapid Assessment of Coral, Fish, and Algae Using the AGRRA Protocol (MSD-00-3)

The Economic Contribution of Whalewatching to Regional Economies: Perspectives From Two National Marine Sanctuaries (MSD-00-2)

Olympic Coast National Marine Sanctuary Area to be Avoided Education and Monitoring Program (MSD-00-1)

Multi-species and Multi-interest Management: an Ecosystem Approach to Market Squid (*Loligo opalescens*) Harvest in California (MSD-99-1)